A MANUAL ON THE HYDRAULIC RAM FOR PUMPING WATER

S.B. Watt

Practical ACTION PUBLISHING

Practical Action Publishing Ltd
27a Albert Street, Rugby,
CV21 2SG, Warwickshire, UK
www.practicalactionpublishing.org

© Intermediate Technology Publications 1975.

First published 1975\Digitised 2013

ISBN 10: 0 90303 115 9
ISBN 13: 9780903031158
ISBN Library Ebook: 9781780441603
Book DOI: http://dx.doi.org/10.3362/9781780441580

Since 1974, Practical Action Publishing (formerly Intermediate Technology Publications and ITDG Publishing) has published and disseminated books and information in support of international development work throughout the world. Practical Action Publishing is a trading name of Practical Action Publishing Ltd (Company Reg. No. 1159018), the wholly owned publishing company of Practical Action. Practical Action Publishing trades only in support of its parent charity objectives and any profits are covenanted back to Practical Action (Charity Reg. No. 247257, Group VAT Registration No. 880 9924 76).

Introduction

The automatic hydraulic ram is a pumping device that has been widely used for nearly a century in rural areas, for lifting water to heights of over 100 metres. It is an ideal machine for water pumping if certain conditions are satisfied, because it works solely on the power from falling water carried in a pipe from a spring, stream or river, without any need for an additional power source. It is completely automatic, and has an exceptional record of trouble free operation. It cannot be used everywhere, however. It cannot be used to pump still water from a well, pond or lake, unless there is a separate, flowing water source nearby.

The simple ram pump described in this manual has been made and used since 1948, but we have only limited information on its performance. This has been taken from the VITA publication (Ref. 6.6 in the bibliography) and our own very limited laboratory tests (Section 5). We would welcome information from anyone who has built one of these ram pumps, operated it over a long period, and who has also measured the pumping rate and surveyed the site conditions.

Although this method of making a ram pump has already been published by VITA (6.6.), we have written the manual to do two things. We hope to demonstrate to field workers how they can build the ram and have the confidence to overcome any difficulties in installation and tuning. We also hope to provide information to those with some technical and workshop experience that will enable them to manufacture larger rams.

In Part 1, we describe how a simple ram pump can be made from commercial pipe fittings, how to choose a site for the ram, how to install and adjust the ram, and the sort of maintenance that the ram pump will need during its working life. We have tried to write the manual in non-technical language so that it can be used by people with little or no technical training.

In Part 11, we describe in greater detail the range and limits of operation of ram pumps, and the different materials that have been used to make them. We have taken information given in the trade literature of a commercial ram manufacturer, Blakes Hydrams Ltd., (Ref. 6.9), to describe the simple calculations that you will need to design a different sized ram pump for different site conditions. We also include in Part 11 the results of our limited laboratory tests on the ram, which describes the tuning procedure.

We have not attempted to describe the complex relationship involved in the hydraulic behaviour of the water moving through the ram, this is very difficult to follow unless you have substantial experience of fluid mechanics. An annotated bibliography listing the main sources of information that we used to write this manual is included in the last section of Part 11.

S.B.Watt.
May, 1975.

Index

PART II   A MORE TECHNICAL LOOK AT HYDRAULIC RAM PUMPS

LIST OF TABLES

LIST OF FIGURES

FIG.1 THE ARRANGEMENT OF A TYPICAL RAM ASSEMBLY

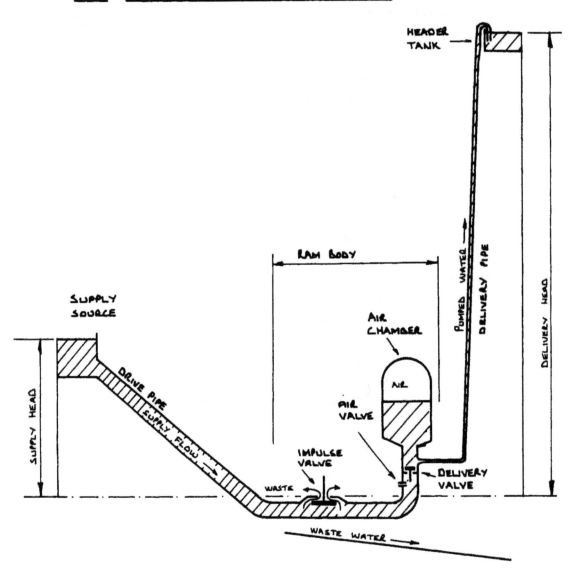

The vertical distance between two water levels is known as the 'head' of water available and is a measure of the water pressure. For instance, the pressure in the ram body when it is full of water and not pumping, is known as the supply head; similarly the pressure in the air chamber with the delivery valve closed, is the delivery head.

Part 1   How to make and Install a a Simple Ram Pump Constructed from Water Pipe Fittings

1. A Description

The automatic hydraulic ram is used for pumping water. It works by pumping a small fraction of the water that flows through it from a supply source, to a level that can be much higher than the source. The ram can only be used in places where there is a steady and reliable supply of water, with a fall sufficient to operate the ram.

The ram described in this manual needs to have a fall of at least 1 metre from the source to the ram, and a flow at the source greater than 5 litres per minute. The amount of water that it can pump to different heights is given in Table 1. (Page 7)

In places where this ram can be used, it has many advantages over other pumps powered by hand, animal, wind, or motors, despite the fact that it wastes a lot of water:-

a) it does not need an additional power source and there are no running costs,

b) it has only two moving parts, and these are very simple and cheap to maintain,

c) it works efficiently over a wide range of flows, provided it is tuned in correctly,

d) it can be made using simple work shop equipment.

2. How it works

A labelled diagram of a typical working ram installation is shown in Fig. 1.

Water flows down the drive pipe from the source and escapes out through the impulse valve. When the flow of water past the impulse valve is fast enough, this flow and the upward force on the valve causes the valve to shut suddenly, halting the column of water in the drive pipe. The momentum of the stopped column of water produces a sudden pressure rise in the ram, which will, if it is large enough, overcome the pressure in the air chamber on the delivery valve, allowing water to flow into the air chamber and then up to the header tank.

The pressure surge or hammer in the ram is partly reduced by the escape of water into the air chamber, and the pressure pulse 'rebounds' back up the drive pipe producing a slight suction in the ram body. This causes the delivery valve to close, preventing the pumped water from flowing back into the ram. The impulse valve drops down, water begins to flow out again, and the cycle is repeated.

A small amount of air enters through the air valve during the suction part of the ram cycle, and passes into the air chamber with each surge of water up through the delivery valve. The air chamber is necessary to even out the drastic pressure changes in the ram, allowing a more steady flow of water to the header tank. The air in the chamber is always compressed, and needs to be constantly replaced as it becomes mixed with the water and lost to the header tank.

The ram is 'tuned' to pump the greatest amount of water possible, and this normally occurs when the ram cycle is repeated or 'beats' about 75 times each minute.

3. Is your site suitable for the ram?

You can install this ram at your site without doing any survey work to measure the flow of water at the source, or the supply and delivery heads at the site, and it will probably work perfectly well. However, it is often necessary to know if the ram is capable of pumping the amount of water you need, or whether you need a larger ram. Measuring for this information is not difficult, and is described below.

3.1 MEASURING THE FLOW OF WATER AT THE SOURCE.

The first thing you must measure, is the flow of water at the source, to see if it is enough to operate the ram; some people with experience can estimate this by eye.

Naturally occuring sources of water tend to dry up during the year, and you must make allowance for this if you use your measurement of water flow to calculate the pumping rate of the ram, otherwise your water supply may be less than you planned for.

a) Measuring a small flow, such as a spring.

When the flow is very small, you can measure it by constructing a temporary dam, and catching the water in a bucket. The amount of water (in litres) that flows into the bucket in one minute can then be measured. The dam may be made from any material, wood, metal sheet, planks etc., but you must make sure that there are no leaks:-

2.

a.) MEASURING A SMALL FLOW.

b) Measuring larger flows.

The ram described in this manual requires only a small amount of water to make it work, and often you can see by looking if the flow is large enough. However, if you are going to make or buy an expensive larger ram, it is essential to know how much water there is to be taken from the source.

Larger flows are measured using a timber plank or ply wood weir, with a 90° V-notch cut into the top:-

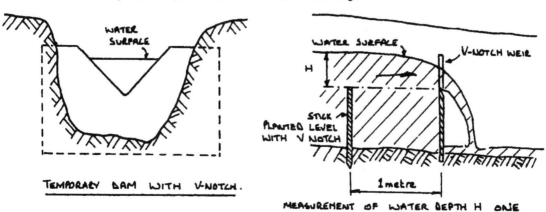

TEMPORARY DAM WITH V-NOTCH.

b) MEASURING LARGER FLOWS

MEASUREMENT OF WATER DEPTH H ONE METER UPSTREAM FROM V-NOTCH WEIR.

The depth of water flowing through the weir is measured about 1 metre upstream of the weir, and you can then use the graph in Figure 2 to read how much water is flowing

Example.

Depth of water measured 1 metre upstream = 10 cms.
From graph, the flow is then read as 275 litres/min.

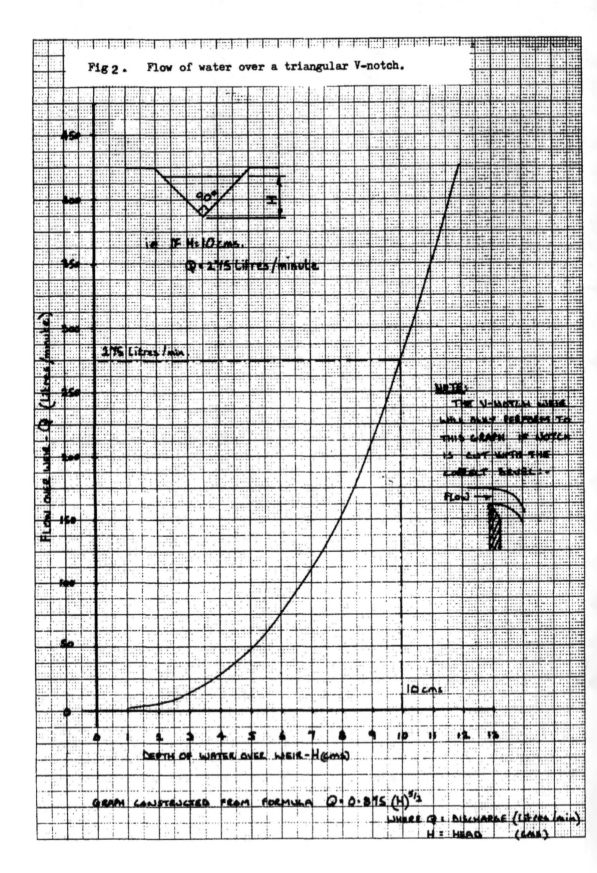

Fig 2. Flow of water over a triangular V-notch.

GRAPH CONSTRUCTED FROM FORMULA $Q = 0.875 (H)^{5/2}$

WHERE Q = DISCHARGE (Litres/min)
H = HEAD (CMS)

The weir must not leak around its sides, and the graph
can only be used if all the water flow is contained within
the notch.

3.2   MEASURING THE SUPPLY AND DELIVERY HEADS.

Most rams will work at their best efficiency if the supply head is about
one third of the delivery head, but often the site will not allow this, and
you must then try to make the supply head as large as possible; this will be
necessary if the source is a slow moving stream or river which has a shallow
slope. The supply head can be increased by leading the water from the supply
source along a feeder canal or pipe to the drive pipe inlet:-

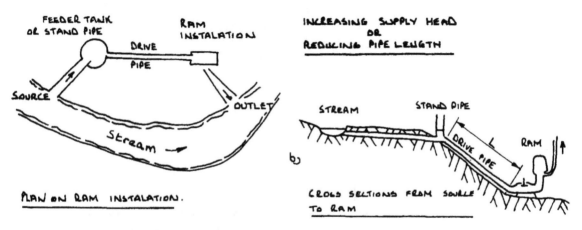

You will only need to measure the supply and delivery heads if you need
to make sure that the ram will pump enough water, or if you have to buy a
larger ram. The flow at the supply source, and the delivery and supply heads,
can be used to calculate how much water this ram will pump. See Table 1. (Page 7)

The differences in level between the source and the ram, and the header
tank and the ram, can be measured using a surveyors dumpy level, a clinometer,
or even a carpenters spririt level attached to a stick. A method of measuring
the supply head is described below:-

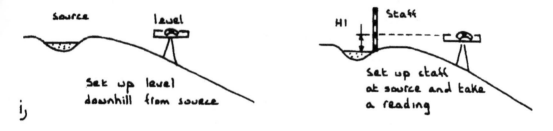

leave level in
same position

move staff
downhill and take
a reading

H2

ii)

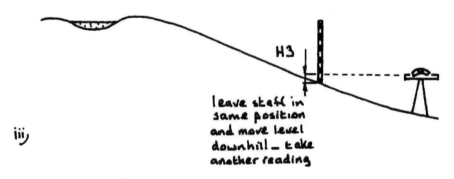

H3

leave staff in
same position
and move level
downhill — take
another reading

iii)

i. Set up the level near the source, and take a reading on a
graduated measuring staff held by your assistant on the
water surface of the source. Record this reading in a
note book. (H1).

ii. Turn the level around on the same spot, and ask your
assistant to carry the measuring staff down hill. The
staff is held upright, and you take a second reading which
you again record. (H2).

iii. Your assistant will stay on the same spot with the measuring staff whilst you carry the level down hill again to a
position below your assistant. Set up the level again,
and repeat stages i and ii above.

You repeat this process until the ram site is reached,
and the supply head can be calculated as follows:-

$$\text{Supply Head} = H_2 - H_1 + H_4 - H_3 + \ldots \text{etc.}$$

The delivery head is measured in a similar way.

4. Designing the ram.

4.1 HOW MUCH WATER CAN THE RAM PUMP

The simple ram pump made from commercial pipe fittings described in Section 5 of this manual, needs a supply flow of at least 5 litres each minute. Using this supply flow, the smallest amounts of water that this ram may be expected to pump each day for different supply and delivery heads, are given in Table 1.

TABLE 1   DAILY PUMPING RATES FOR RAM PUMP (litres of water)

Supply Head (metres)	Delivery Head (metres)										
	5	7.5	10	15	20	30	40	50	60	80	100
1	400	200	150	80	70	50	30	20			
2		550	390	250	200	130	80	60	50	30	
3			650	450	320	220	150	130	100	70	40
4				650	430	300	200	150	130	90	60
5				750	550	370	300	250	200	120	90
6					700	450	350	300	250	150	120
7						550	410	320	270	200	150
8							450	370	300	250	150
10							600	450	400	300	230
12							750	550	470	350	280
14								650	550	400	330
16									620	470	370
18									700	520	420
20										600	450

We have not been able to test the ram pump over this wide range of supply and delivery heads. We have assumed that it will pump at only one half the rate of a comparable commercial ram manufactured by Blakes Hydrams Ltd. (see Table 2, page 30).

The ram will pump at a faster rate if the impulse valve is properly tuned, or if the supply flow is more than 5 litres per minute.If for example, your ram installation can be tuned to allow a flow of 15 litres per minute down the supply pipe, then the ram will pump three times the amount given in Table 1.

The greatest amount of water this ram can use from the source is governed by the size of the ram itself and if the ram installation is to use more water (and therefore be able to pump more water), then a larger ram should be chosen. How to choose the correct ram size is given in Part 11 of this manual.

If you find that your ram installation is not large enough to pump the amount of water you need, you can construct a duplicate ram alongside the original ram. The drive pipes should be separate, but you may use the same delivery pipe. Some installation have batteries of small rams, often 5 or more, next to each other.

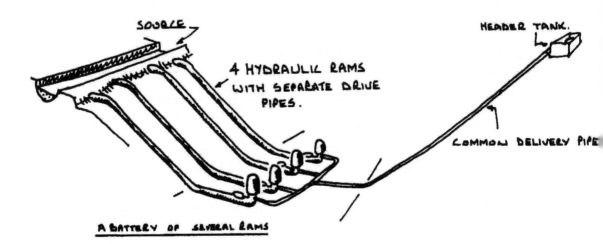

A BATTERY OF SEVERAL RAMS

4.2 CHOOSING THE SIZE OF THE DRIVE PIPE.

The drive pipe is really the most important part of the ram installation - it carries the water from the source to the ram, and contains the pressure surge of the water hammer. It must be made from good quality steel or iron water pipe - plastic and concrete pipes are useless for drive pipes.

The diameter and length of the drive pipe is very important, although the ram will work satisfactorily if the ratios of pipe length (L) to diameter (D) are between the limits $\frac{L}{D}$ = 150 to 1000. These are very broad limits. We suggest that you try to install a drive pipe with an $\frac{L}{D}$ ratio of 500, or choose a length that is four (4) times the supply head, whichever is the smaller. The theory behind the drive pipe is described in greater detail in Part 11.

Example

Supply Head   =   4.0 metres
Drive pipe diameter (D) =   25 mm.

8.

a)   Use $\frac{L}{D}$ = 500

      L   = 500 x 25 = 12500 mm or <u>12.5 metres</u>.

b)   Use L   = 4 x Supply head

      L   = 4 x 4.0 = <u>16.0 metres</u>.

The ram will work equally well if the drive pipe is cut from 25 mm pipe at either of these lengths, and you should choose the length which is most convenient for your site.

4.3 <u>CHOOSING THE DELIVERY PIPE SIZE</u>

Unlike the drive pipe, you can make the delivery pipe from any material, provided it can stand the pressure of water leading up to the delivery tank. The delivery pipe should have an internal bore of 20 mm; plastic hose pipe is quite satisfactory if it is strong enough.

The water from the ram can be pumped for great distances provided that the delivery head is small enough; in this case, the ram has to spend effort forcing water through the pipe, and you should try to keep the delivery pipe fairly short.

4.4 <u>CHOOSING THE SIZE OF THE HEADER TANK</u>.

One of the great advantages of a ram pump is that it works automatically and continuously, which means that it is always pumping water to the header tank.

If you think about the way that you use water in your household, you will see that during certain periods of the day, you will need a relatively large amount from the header tank. At other times, during the night, for instance, you will most likely use very little water.

The header tank must therefore be large enough to hold enough water in reserve to supply your needs during periods of peak demand.

Even when you choose a header tank of correct size, there will be times when it overflows. You should therefore fit an overflow pipe to the tank, and lead the waste water to your garden or fish tank.

The way to choose the tank size is to estimate your daily water requirements, and make your tank to contain half this amount. If you find the tank is too small, you can easily add a second tank.

5. Building the ram.

You can build a ram from any size of pipe fittings that you have available, and the way that these will work is described in Part 11. The ram described here has a drive pipe bore of 30 mm. The ram body is made from pipe fittings of 50 mm internal bore, so that the impulse and delivery valves can have large openings: the relatively small sizes of commercial pipe fittings are a major disadvantage for ram construction, and effectively limit the maximum ram size that can be made. The finished ram is shown below and in Fig.3

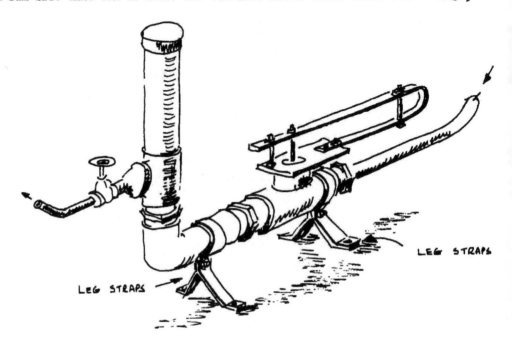

The main points you should note when you intend to build this ram are:

   a) the capacity of the ram depends on the size of the
       impulse valve which allows the water to discharge.
       The pipe fittings are therefore several sizes larger
       than the drive pipe.

   b) the flow of water through the ram should not be
       restricted by sharp changes of direction of water
       flow or by the sudden junction of different sized
       pipes.

   c) the ram experiences savage pounding during its working
       life and all the parts, connections and valves must be
       strong enough to stand the stresses.

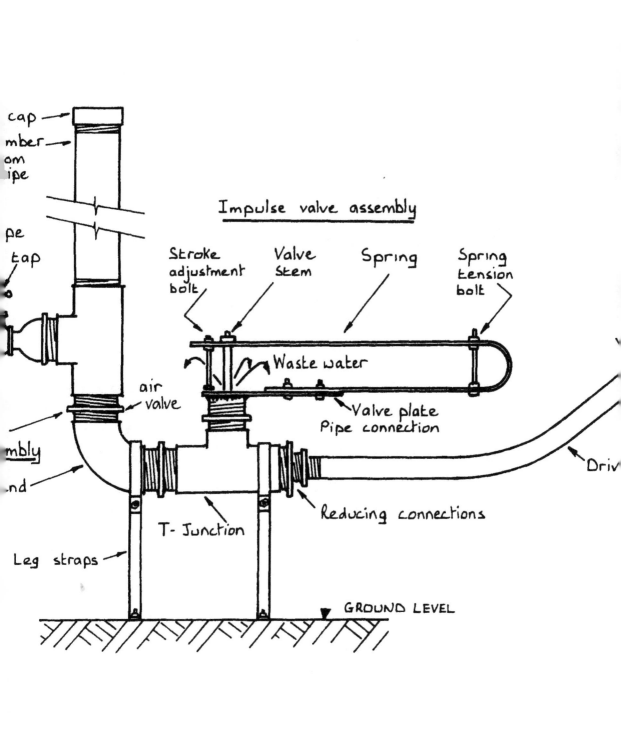

cap

mber

om
ipe

pe

tap

Impulse valve assembly

Stroke
adjustment
bolt

Valve
Stem

Spring

Spring
tension
bolt

Waste water

air
valve

Valve plate
Pipe connection

mbly

nd

Reducing connections

T-Junction

Driv

Leg straps

GROUND LEVEL

d) there are obviously any number of combinations of
   pipe fittings which can make up a ram body, and
   the one described below can be modified to suit
   available fittings.

5.1 MAKING THE IMPULSE VALVE.

Weld or braze a 50 mm threaded pipe junction onto the valve plate shown
in Fig.4.1 centrally over the 30 mm diameter hole:-

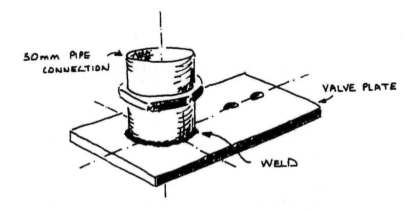

WELDING CONNECTOR TO PLATE

This will leave a lip inside the pipe connection about 10 mm wide all
round, which will act as a seating for the impulse valve washer. File or rub
and smooth the valve plate over the valve seating area to prevent wear on the
valve washer. The two elongated holes, each 6 mm diameter on the valve plate,
are to hold the valve spring.

The valve spring is made from a strip of mild steel, 650 mm long,
30 x 2 mm in cross section, marked out and drilled as shown in Fig.4.3. Bend
the spring to shape around a 50 mm pipe, with the bend centre line on the strip
in the position as shown below, this will set the spring with the drilled holes
in the correct positions:-

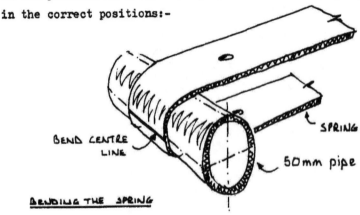

BENDING THE SPRING

Bolt the spring onto the valve plate, which has elongated holes to allow the impulse valve stem to be adjusted for correct seating.

The impulse valve itself is made up from a 6 mm diameter bolt, tube and washers which you assemble through the valve plate to the valve spring, Fig. 4.4

Finally, add spring tension and the valve stroke adjusting bolts. These allow the ram to be tuned for maximum efficiency. You can see that the impulse valve assembly can be removed from the ram for maintenance by just unbolting the spring, then unscrewing the pipe connector and valve plate:-

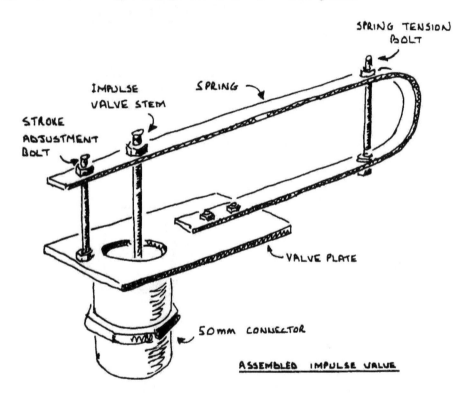

ASSEMBLED IMPULSE VALVE

We have chosen this system of impulse valve assembly because it has no wearing parts except for the valve rubber. It is possible that with time, the valve spring will work harden and break; it is also possible that the spring assembly will be damaged during floods if the ram is installed on the side of a stream. An alternative more robust design for the impulse valve assembly is described below.

5.1 B AN ALTERNATIVE IMPULSE VALVE.

The impulse valve assembly described above has been taken from a design by VITA, and as far as we know, it works quite satisfactorily.

<u>FIG 4</u>   <u>CONSTRUCTION OF IMPULSE VALVE</u>   (Dimensions in millimetres)

4.1.   VALVE PLATE,   150 x 80 x 3 mm   Mild steel plate

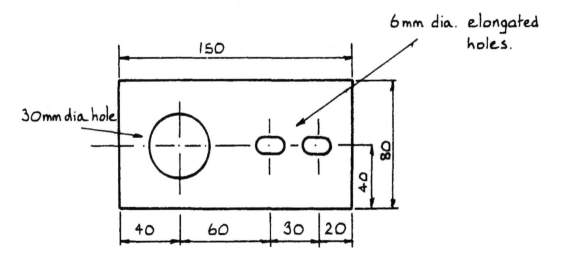

4.2.   PIPE CONNECTION,   50 mm diameter, mole threaded

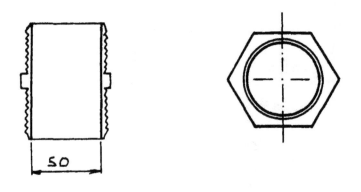

4.3.   SPRING, 650 x 30 x 2 mm   Mild steel strips

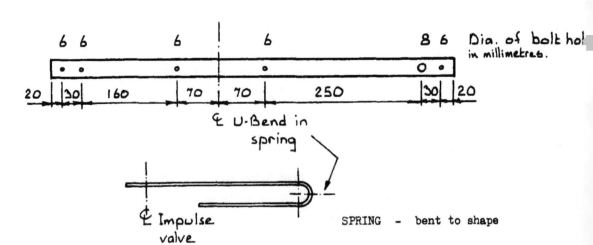

4.4 IMPULSE VALVE STEM

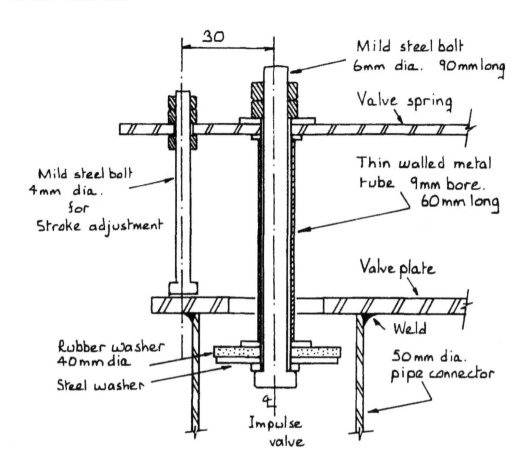

30

Mild steel bolt
6mm dia. 90mm long

Valve spring

Mild steel bolt
4mm dia.
for
Stroke adjustment

Thin walled metal
tube 9mm bore.
60mm long

Valve plate

Rubber washer
40mm dia

Steel washer

Weld

50mm dia.
pipe connector

4
Impulse
valve

4.5. SPRING TENSION BOLT

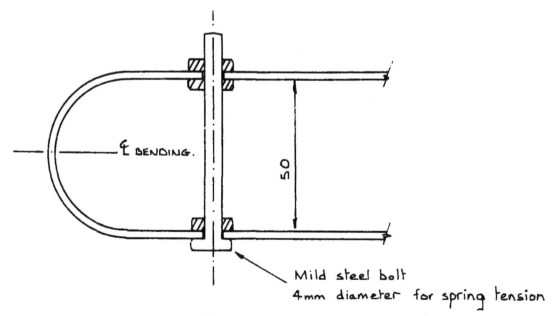

₵ BENDING.

50

Mild steel bolt
4mm diameter for spring tension

We include in this section a more robust impulse valve with a sliding valve which will wear in time. The impulse valve in this case works by falling under its own weight at the finish of each ram cycle:-

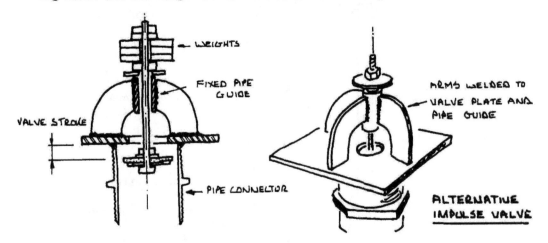

The valve stem is fitted through a fixed pipe guide supported above the valve plate by arms welded both to the pipe and valve plate. The pipe connector is welded as before over the centre of the 30 mm diameter hole in the valve plate.

Choose the pipe guide and the valve stem bolt so that they have a close but easy fit. Alternatively, the pipe guide can be chosen to hold a replaceable brass or plastic sleeve which will take the wear from the moving valve stem bolt.

The valve stroke is set by adjusting the nuts on the top of the valve stem bolt, and the weight of the valve can be altered by adding weights onto the bolt.

We have not built or tried this impulse valve assembly, but there is no reason why it should not work. Tuning the ram will be a similar process to that described in Section 7.

5.2 MAKING THE DELIVERY VALVE.

The delivery valve prevents the pumped water from flowing back into the ram after the pressure pulse has been dissipated. It is therefore a non return valve, and you can make it very simply by welding or brazing a cut and drilled piece of 3 mm steel plate into the top of a 50 mm pipe connector:-

FIG.5 CONSTRUCTION OF DELIVERY VALVE.

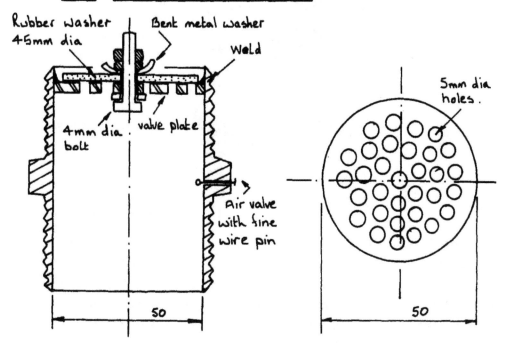

Rubber washer 45mm dia

Bent metal washer

Weld

4mm dia bolt

valve plate

Air valve with fine wire pin

50

5mm dia holes.

50

Take care when the valve plate is welded or brazed into the pipe connector that the plate remains clean, otherwise the valve rubber will not seat correctly and the valve will leak.

Non Return valve from 3mm steel plate —

drilled with 5mm dia holes and polished smooth. Larger holes may cause the valve rubber to distort and leak.

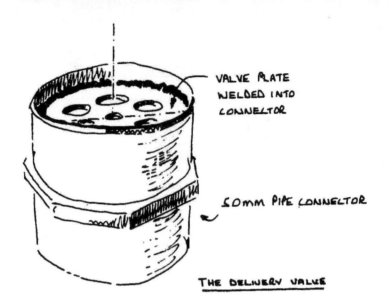

VALVE PLATE
WELDED INTO
CONNECTOR

50mm PIPE CONNECTOR

THE DELIVERY VALVE

Cut the plate to shape and file smooth to fit exactly into the end of the pipe connector, and weld or braze it into place. Attach a rubber washer to the plate and bolt it into position; the washer must be flexible enough to allow water to pass easily, but must be firm enough to support the water pressure from the air chamber. The cupped washer above the rubber valve holds the valve in place.

The air valve is made simply by drilling a small hole 1.0 mm in diameter in the side of the pipe connector and below the delivery valve. This is partially blocked by a fine wire split pin which moves with pressure changes in the ram, keeping the hole open and allowing air to enter. Fig.5.

Make sure on assembly that the air valve is placed on the opposite side to the delivery pipe outlet, otherwise the air entering the air chamber is likely to escape into the delivery pipe; it is, of course, essential that the air feeder valve is located below the delivery valve.

5.3 MAKING THE AIR CHAMBER.

Cut a 1 metre length of 50 mm diameter water pipe, and thread each end. Screw one end into the delivery pipe T-junction pipe fitting, and seal the top with a cap.

5.4 MAKING THE MOUNTING LEGS.

Make the mounting legs from any available scrap strip iron, and drill, bend, and bolt these around the ram body. The legs can be bolted to the ground when the ram is assembled at the site if you want the ram to be a permanent fixture.

6. Assembling the ram at the site.

a) Assemble the pipe fittings using plenty of pipe joint compound. Screw these firmly together and adjust them for the correct position in the ram assembly. They must be completely free from leaks.

b) The impulse and delivery valves must move freely and when closed seat evenly on the valve plates.

c) Set the ram level on the mounting legs at the required site, and attach the drive and delivery pipes. Flush these pipes with clean water before connection.

d) The drive pipe should be laid as straight as possible with no sharp bends, and it should have no upward kinks which will trap air.

e) The inlet to the drive pipe must always be submerged, or air will enter the pipe and prevent the ram from working.

7. Tuning the ram.

The ram should be tuned to pump the greatest amount of water to the delivery tank. Tuning is not difficult, and you will find that the ram will pump some water at most settings of the impulse valve assembly.

The amount of water that the ram will pump, and the number of valve beats each minute, are measured for different valve settings, and the results compared to find the best setting for the ram. You can do this quite easily:-

a) Hold the impulse valve closed, and adjust the 'stroke adjustment bolt' (SAB) until there is a gap of about 15 mm between this bolt and the valve plate. This can most simply be done by slipping a measured pile of steel washers under the bolt and screwing the bolt down onto them.

b) Remove the washers, release the impulse valve, and adjust the 'spring tension bolt' (STB) until the SAB just touches the valve plate. Shortening the STB will bend the spring down.

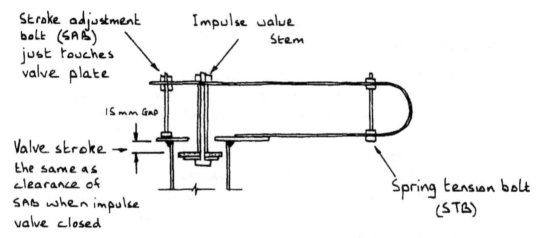

Stroke adjustment bolt (SAB) just touches valve plate

Impulse valve stem

15 mm Gap

Valve stroke — the same as clearance of SAB when impulse valve closed

Spring tension bolt (STB)

c) Nip tight the STB and SAB nuts, and allow water to enter the drive pipe. Hold the impulse valve closed until the drive pipe is full of water, then release the valve, moving it up and down by hand several times. The ram should now work by itself.

d) If the valve stays open allowing water to flow out, the spring is too tight, and you should stop the flow of water, and reset the SAB and STB in the way described in a, b, and c, above to give a stroke of 13 mm.

e) When the valve works correctly, repeat a, b, and c, above for valve stroke settings of 13, 11, 9, 7, 5, 3, millimetres, measuring for each setting the amount of water that is pumped and the valve beats each minute.

f) Compare the pumping rates, and reset the STB and SAB as described in a, b, and c, to the stroke setting that gave the best pumping rate. If the pumping rates for several of the valve settings are similar, choose the setting with the smallest stroke - this will mean a smaller spring tension and therefore less wear.

g) The results of our experiments on one of these rams are given in Part 11 of this manual. We obtained the best pumping rate from an initial valve stroke setting giving a valve beat of 100 cycles per minute, by tightening the spring tension bolt until the valve beat was 75 cycles per minute. The ram you make will work in a different way to ours, and you will have to fiddle with the impulse valve to find the best setting.

8. What to do if the ram doesn't work.

There are only two moving parts in an automatic hydraulic ram, and there is very little that can go wrong. However, possible causes of failure are listed below:-

a) Impulse valve does not work.

Check seating of valve washer on valve plate; the valve should not leak when held closed, and should not catch on the side of the pipe connector.

Check to see if there is any debris or obstruction in the drive pipe or ram body.

b) Delivery valve does not operate as a non-return valve.

This can be seen if the water level in the delivery pipe surges during operation, or falls when the ram is not working. The valve should be cleaned and checked for wear.

c) Ram pumps too much air.

Check air feeder valve; if it is too big it will allow large volumes of air to enter the ram, and a larger wire split pin should be used.

Check that air does not enter the ram through loose joints; the joints should be well sealed with pipe compound.

Check that inlet to drive pipe is submerged, otherwise air will enter drive pipe, spoiling the performance of the water hammer.

d) Ram pumps with a loud metallic sound.

Check that air feeder valve is working to allow enough air to enter below the non-return valve; a small spurt of water should come from this valve with each cycle. If there is not enough air entering the ram air chamber, fit a smaller split pin.

Check that air feeder valve is on the opposite side to the delivery pipe, or the air will be pumped with the water directly to the header tank.

Check that there are no air leaks from air chamber due to bad pipe fitting.

9. Maintenance of the ram after installation.

9.1 THE SUPPLY SOURCE

It is obviously essential to prevent dirt from entering the drive pipe or leaves from blocking its entry. So it may be necessary to provide a grating at the off-take from the river or stream supplying the water in order to keep back floating leaves, and a sump should be provided at the feeder tank to collect silt.

9.2 MAINTENANCE TASKS

Maintenance involves keeping gratings and filters clear, and cleaning the feeder tank and sump, as well as caring for the ram itself. The maintenance tasks which you must carry out are likely to be as follows:

(a) dismantling the ram to remove dirt,

(b) clearing air locks in the pipe system,

(c) adjusting the tuning; tightening bolts which work loose,

(d) changing the valve rubber; adjusting the seating of valves,

(e) keeping the inflow to the drive pipe free of debris; clearing filters and gratings.

9.3 FREQUENCY OF MAINTENANCE

Rams have an exceptionally good reputation for trouble-free running, and maintenance will probably not need to be very frequent. The way in which the necessary maintenance is arranged, and the question of whether this type of ram is suitable for a particular application, depends very much on who is available to carry out the maintenance. Is there somebody living locally who can have a look at the ram at least once every week, or is there a technician from somewhere else who can come only at intervals of several weeks?

Tuning, and the adjustment of valves and bolts, may need to be done more frequently with this particular ram than with some commercial models made from purpose-designed alloys and components; and the need for maintenance may become greater as the delivery head becomes greater. On the other hand, specialised tools and spare parts may be needed for the maintenance of a commercially-built pump. So in general, this ram is best suited to a situation where the person responsible for maintenance lives nearby, and where the delivery head is not too great. A commercial pump may be the best choice when maintenance is done at longer intervals by somebody with access to a wide range of tools and components.

Part II A more Technical look at Automatic Hydraulic Rams

1. Introduction

This part of the manual will be of interest to those who have a basic understanding of engineering materials, and fluid mechanics. It will be of use to those who wish to build rams with different sizes to the one described in Part 1.

Commercially available rams have been redesigned and refined by field experience until they work well under all conditions with the minimum of maintenance. They are made from solid iron castings, and are extremely robust - some ram installations have been working for nearly 100 years.

The size of the ram described in Part 1 is necessarily limited by the size of the available pipe fittings. The strength of the pipe fittings also limits the size of the ram, - it is doubtful if pipe fittings would stand up to the savage loads experienced by commercial rams under conditions of high supply heads and supply flows.

Large rams must therefore be made from iron castings, or welded steel pipe. A technical description of casting and workshop processes are outside the scope of this manual, but if you have these skills and the necessary equipment available, this part of the manual will show you the main design features to be considered before you design a ram for production.

2. Ram performance

The way that automatic hydraulic rams work is outlined briefly in Part 1, and there is little to add to this except to show on a diagram one pressure pulse cycle of a ram. The diagrams in Fig. 6 show in a very simplified and ideal form the pressure and velocity at the end of the drive pipe, and the position of the impulse valve, during one cycle.

Rams were built and used for nearly a century before any intensive research was carried out on their operational characteristics, and they seem to be almost foolproof in operation. Recent research has clarified the way that rams seem to work, - references to this work can be found in the bibliography.

FIG. 6   DIAGRAMS SHOWING ONE PRESSURE PULSE CYCLE OF HYDRAULIC RAM.

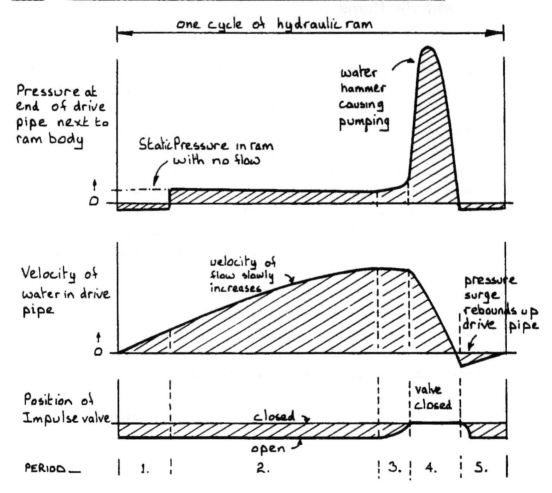

Period 1.   End of previous ram cycle, velocity of water through ram begins to increase through open impulse valve; slight negative pressure in ram body.

Period 2.   Flow increases to a maximum through open impulse valve.

Period 3.   Impulse valve begins to close causing pressure to build up inside ram.   The velocity of flow through the ram has reached a maximum, the maximum velocity being controlled by ram size.

Period 4.   Impulse valve has closed, causing the pressure pulse or hammer to pump some water through the delivery valve.   Velocity of flow through ram rapidly decreases.

Period 5.   Pressure pulse rebounds back up drive pipe, causing a slight suction in ram body.   Impulse valve opens under this suction and its own weight.

Water begins to flow again through the impulse valve, and the ram cycle is repeated.

3. Some design considerations

3.1 CONSTRUCTION MATERIALS

When the column of water in the drive pipe is suddenly retarded by the closing impulse valve, the pressure build up compresses the water, causing the elastic materials making up the drive pipe and ram body to stretch. In this way, part of the energy of the pressure pulse is used in straining the pipe walls. An ideal installation would be one made of completely rigid, inelastic, materials, and in this case an instantaneous reduction in the flow velocity of 0.3 metres/second would cause a pressure head of about 4.6 kgm/cm^2 i.e. a head of nearly 48 metres of water. With the materials and valves that are available, it is not possible to achieve this and Fig.7 shows the equivalent pressure head produced when different materials are used to construct the ram.

3.2 DRIVE PIPE

a) Length and diameter

Research has shown that the size of the drive pipe does not affect the ram performance over a very wide range of flow conditions, and the pipe diameter is usually determined by the pipe materials available. It is not possible to calculate the size of pipe needed - the flow of water down the pipe varies cyclically, and the amount of water that the ram will use depends mainly on the size of the impulse valve and the supply and delivery heads.

Some idea of a suitable drive pipe diameter for a ram of known size can be found from the information given in Table 2 on commercial rams. The ratio of pipe length to diameter ($\frac{L}{D}$) should in any case be between the limits $\frac{L}{D}$ = 150 to 1000; outside these limits, the performance of the ram is impaired. These limits seem to be determined by the ability of the water column in the drive pipe to accelerate after it has been stopped.

The cost of the drive pipe is a very major item in a ram installation, and the pipe should therefore be chosen to have a small diameter; however, if the pipe is too small in diameter, the viscosity of the water and the friction on the pipe walls will slow down the accelerating water column and reduce the efficiency of the ram.

The drive pipe diameter is first chosen to correspond with the size of the ram body, or from a comparison with the commercial rams, and the length of the pipe found from this, using a ratio $\frac{L}{D}$ of about 500.

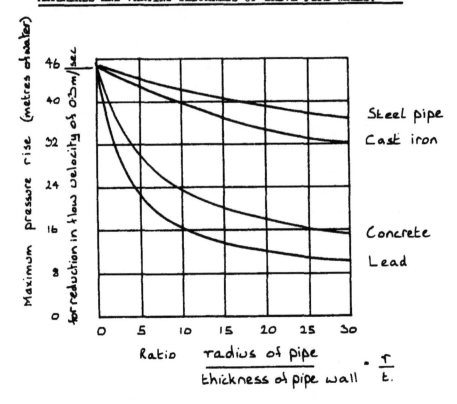

An ideal drive pipe would be made from steel, and the walls of the
pipe would be very thick in relation to the pipe diameter. The diagram
above shows that with an increase in the   ratio $\frac{r}{t}$, the maximum
pressure increase can be expected to fall and the ram will not be able
to pump very efficiently. It can also be seen that concrete is a poor
material for ram construction.

If the instantaneous reduction in flow velocity is 1 metre/second,
then the maximum pressure head increase will be $\frac{1.0}{0.3}$ x 46 or 154 metres of
water.

With an instantaneous reduction in flow velocity of 5 metres/second,
the maximum pressure head should be $\frac{5.0}{0.3}$ x 46, or 765 metres of water. In
practice, the ram would have to be very large to allow water to reach a
velocity of 5 metres/second down the drive pipe.

Some ram manufacturers suggest that the drive pipe length be 4 or 5 times the supply head.

The length of the drive pipe is quite an important dimension for the ram design - the compression wave of water must reach the open source and be dissipated before the water in the drive pipe can flow again through the impulse valve. The drive pipe length would be critical for a site which had a source a long way from the ram, with a low supply head. In this case, a stand pipe or feeder tank should be installed.

The inlet to the drive pipe must always be submerged to prevent air from entering the pipe; air bubbles in the drive pipe will absorb the energy of the pressure pulse, reducing the ram efficiency. For this reason, the drive pipe must not be laid with any upward bends or humps that could act as air traps.

b) Pipe smoothness.

The column of water in the drive pipe accelerates and is stopped very rapidly many times a minute. The walls of the drive pipe should therefore be as smooth as possible, otherwise the efficiency of the ram will be greatly reduced. This is especially true if a small diameter drive pipe is chosen; a large diameter drive pipe will have much lower velocities and smaller friction losses. Concrete lined pipes give a rough wall finish with fairly high friction losses.

3.3 IMPULSE VALVE

This is a vital part of the ram, and it should be designed so that it's weight and stroke can be adjusted for tuning:

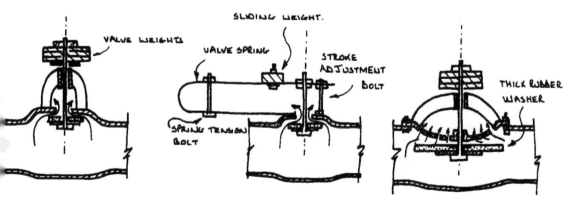

a) SIMPLE CLACK VALVE   b) CLACK VALVE WITH SPRING   c) FLEXIBLE RUBBER WASHER.

SOME TYPES OF SUCCESSFULL IMPULSE VALVES.

A heavy weight and a long stroke will allow high flow rates through the impulse valve, building up the powerful hammer pulse needed to drive water to high heads; a small weight and short stroke will 'beat' more quickly and deliver larger volumes to lower heads. There has been very little research carried out into the best shape of the impulse valve, but the simple clack valve seems to perform quite efficiently.

Various spring devices have been tried to cause the impulse valve to shut and open more quickly, and several commercial models incorporate these refinements, (see b above). It is not known if this increases the efficiency of the ram to any great extent, but it does avoid the need for sliding bearings which have to be replaced when worn.

3.4 DELIVERY VALVE

The delivery valve should have a large opening to allow the pumped water to enter the air chamber with little obstruction to flow. It can be a simple non-return valve made from stiff rubber, or operate as a clack valve:

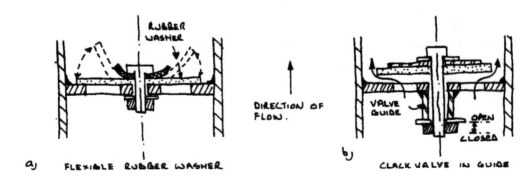

a) FLEXIBLE RUBBER WASHER   b) CLACK VALVE IN GUIDE

DELIVERY NON-RETURN VALVES

3.5 AIR CHAMBER

This should be as large as possible to compress and cushion the pressure pulse from the ram cycle, allowing a more steady flow through the delivery pipe with less friction loss. If the air chamber becomes filled completely with water, the ram will pound savagely and may result in breakage; when this occurs, the ram must be stopped immediately. Some authorities suggest that the volume of the air chamber should be equal to the volume of water in the delivery pipe. On long delivery pipe lengths, this would give an absurdly large air chamber, and a smaller size should be chosen.

3.6 AIR VALVE

The air stored in the air chamber is either slowly absorbed by the turbulence of the water entering through the delivery valve, or is lost into the delivery pipe. This air has to be replaced by the air valve.

The air valve should be adjusted so that it gives a small spurt of water with each compression pulse. If the valve is open too far, the air chamber will fill with air, and the ram will then pump only air. If the valve is not sufficiently open and does not allow enough air to enter, the ram will pound with a metallic sound and break - this condition should be corrected immediately by increasing the opening of the air valve.

PIN MOVES IN AND OUT WITH EACH SQUIRT OF WATER KEEPING HOLE CLEAN

VALVE CAN BE UNSCREWED AND REPLACED.

a) SIMPLE VALVE WITH SPLIT PIN   b) RE-PLACEABLE VALVE

3.7 DELIVERY PIPE FROM RAM TO HEADER TANK

Water can be pumped by a ram to any distance, but a long pipe will involve some work by the ram in moving the water against pipe friction. The delivery pipe may be made from any material, including plastic hose-pipe, providing that it can stand the pressure of the water. Several rams can also be connected to the same delivery pipe if the initial ram install-ation proves to be too small.

The ram should therefore, be located as near as possible to the header tank and the delivery pipe should be made larger with long distances, or with increased volumes of pumped water:-

Pumped water (1000 litres/day)	3	9	14	23	55	90	135
Delivery pipe bore (cms)	2.0	2.5	3.0	4.0	5.0	6.0	8.0

TABLE 2 PUMPING PERFORMANCE OF BLAKES RAMS.

This table gives the quantity, in litres, of water raised every 24 hours, for each litre of supply flow used per minute, under the chosen conditions of delivery head and supply head. These figures have been obtained from field trials on Blakes rams, which operate at efficiencies of about 65%.

Supply Head (H_s) (Metres)	Delivery Head (h_d). (Metres)											
	5	7.5	10	15	20	30	40	50	60	80	100	125
1.0	144	77	65	33	29	19.5	12.5					
2.0		220	156	105	79	53	33	25	19.5	12.5		
3.0			260	180	130	87	65	51	40	27	17.5	12
4.0				255	173	115	86	69	53	36	23	16
6.0					282	185	140	112	93.5	64.5	47.5	34.5
7.0						216	163	130	109	82	60	48
8.0							187	149	125	94	69	55
9.0							212	168	140	105	84	62
10.0							245	187	156	117	93	69
12.0							295	225	187	140	113	83
14.0								265	218	167	132	97
16.0									250	187	150	110
18.0									280	210	169	124
20.0										237	188	140

TABLE 3 CAPACITY OF BLAKES RAMS

This table shows the supply discharge Q_s which can be used by Blakes Hydrams of different sizes.

Size of Hydram (Blakes)		1	2	3	$3\frac{1}{2}$	4	5	6
Internal Diameter	mm	32	38	51	63.5	76	101	127
(bore)	ins	$1\frac{1}{4}$	$1\frac{1}{2}$	2	$2\frac{1}{2}$	3	4	5
Supply Discharge Q_s	From	7	12	27	45	68	136	180
(litres/min)	to*	16	25	55	96	137	270	410
Maximum height to which Hydram will pump water (h_d)	metres	150	150	120	120	120	105	105

* Note: The higher values of Q_s are the volumes of water used by the Hydrams at their maximum efficiency; the rams do not have the capacity to pass larger amounts than those given.

Designing the ram size

The site conditions of supply and delivery head, and supply discharge must first be measured before a ram size can be chosen to pump water at the required rate.

The ram performance data given in Tables 2 and 3 has been obtained by 'Blakes Hydrams Ltd.' (see bibliography) from field trials on their rams. These operate at a maximum overall efficiency of about 65%.

Efficiency $E = \dfrac{q_d \times h_d}{Q_s \times H_s} \times 100\%$ where q_d = pumping rate   (litres/min)
Q_s = supply flow rate (litres/min)
h_d = delivery head   (metres)
H_s = supply head   (metres)

The overall pumping efficiency of a ram depends on the materials used to make the ram, the design of the ram, and its tuning - the efficiency cannot be calculated from basic principles. However, if a ram is designed to the general recommendations given in this manual, its efficiency of operation will not be very much less than that quoted by Blakes Hydrams Ltd. for their equipment, and Tables 2 and 3 may be used with confidence.

<u>Example of ram design calculation</u>

Site measurements:

Supply Head (H_s)   = 5.0 metres
Delivery Head (h_d) = 40.0 metres
Amount to be
pumped/day (q_d)   = 8500 litres

From Table 2, with H_s = 5.0 m, h_d = 40.0 m, and if the flow rate down the drive pipe is 1 litre/min, then 118 litres of water will be pumped each day to the header tank.

But 8500 litres/day are needed at the header tank, and the ram which can pump this needs to be able to use a supply flow of:-

$$Q_s = \frac{8500}{118} = 72.0 \text{ litres/min}$$

Now, from Table 3, a Blakes No. $3\frac{1}{2}$ ram will be satisfactory, or a ram with an internal bore greater than 63.5 mm.

5. Laboratory tests carried out on the ram built from Part 1 instructions.

The ram pump described in Part 1 of this manual, was constructed and tested in the laboratory. Our observations are given below.

5.1 OPERATING THE RAM.

The ram can be made to operate for different conditions:-

a) to pump as much water (q_d) as possible up to the header tank. This means that the flow at the source must always be greater than the flow into the drive pipe.

b) to operate with a limited flow from the source. The pump must then work with this flow (Q_s) at the highest efficiency possible. The water level at the source should always cover the inlet to the drive pipe, or the ram will suck in air and cease to work.

5.2 OBSERVATIONS ON RAM BEHAVIOUR.

The following general remarks may be made:-

a) the pumping rate (q_d) reached a peak during adjustment of valve stroke and spring tension; tuning the ram with the stroke

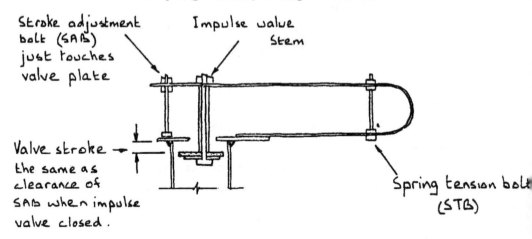

Stroke adjustment bolt (SAB) just touches valve plate

Impulse valve Stem

Valve stroke → the same as clearance of SAB when impulse valve closed.

Spring tension bolt (STB)

adjustment bolt (SAB), and spring tension bolt (STB) is quite critical but is very simple in practice.

b) there are several positions of both the SAB and STB which give the same pumping rate (q_d), but using different amounts of water from the source (Q_s). The setting with the shortest stroke, the lowest spring tension, and using the smallest amount of water, should be chosen in order to reduce valve wear and wastage of water.

32.

c) adding weights to the valve spring will reduce the spring
    tension needed for slow beating on large stroke lengths;
    this may be necessary or the spring will become distorted
    when tuning for high supply heads.

5.3 TESTING THE RAM.

The ram was constructed in the way described in Part 1 and tested in the
laboratory for a wide range of valve settings. The supply and delivery heads
were kept constant throughout the tests, and our results may well form a
different pattern to those of a ram tested under different heads.

The impulse valve was tuned in the way described above, and the pumping
rate (q_d), supply flow (Q_s), spring tension (W), valve beat, and stroke (S),
measured and recorded. The spring tension (W) was the force required to just
hold the valve closed with no water flowing; it was measured using a small
spring balance attached to the impulse valve bolt.

Other tests were carried out on the ram in a similar way by adding weights
onto the impulse valve bolt and taking readings as the valve stroke was reduced.
The STB was not adjusted, and the valve stroke was not measured. These are tests
B1, B2 and B3 in Table 4.

5.4 TEST RESULTS

The results of our tests are given in Table 4 and the following observations can be made:-

a) At a valve stroke of 11 mm the force required to shut the valve
    was 0.91 kgm, and the valve did not shut by itself when water
    flowed.

b) At each valve stroke setting, increasing the spring tension
    increased the pressure of water needed from the flow to shut
    the valve. The supply flow (Q_s) and valve beat varied with
    spring tension, giving different pumping rates (q_d).
    The peak pumping rate does not necessarily mean that
    the ram is operating at its greatest efficiency (E).

c) Decreasing the valve stroke decreased the amount of
    flow through the ram at the initial STB setting.

d) Adding weights to the impulse valve beat does not
    seem to improve the performance of the ram, except
    that it needs less water down the drive pipe. It is
    possible that using weights instead of tensioning the
    spring will lengthen the life of the spring.

Table 4. **Test results on automatic hydraulic ram.**

Supply Head (H_s) = 1.70m   Delivery Head (h_d) = 4.04m

Efficiency E $= \dfrac{q_d \times h_d}{Q_s \times H_s} \times 100\%$

Test No.	Spring Tension W (Kgms)	Valve Stroke (mm)	Valve Beats (cycles/min)	q_d (L/min)	Q_s (L/min)	E%
A1	0.91	11.00	-	-	-	-
A2	0.78	9.5	58	2.60	21.10	29
	0.82	9.5	58	2.70	19.70	32
	0.96	9.5	50	2.40	22.60	25
A3	0.64	8.0	80	2.50	15.64	38
	0.77	8.0	66	2.80	18.02	37
	0.87	8.0	45	1.90	23.70	19
A4	0.54	6.0	96	2.35	10.85	52
	0.64	6.0	78	2.82	14.66	46
	0.82	6.0	58	2.60	19.06	32
	0.95	6.0	48	2.40	23.44	24
A5	0.36	3.5	160	1.32	6.46	49
	0.45	3.5	134	1.76	7.40	57
	0.59	3.5	116	2.00	8.50	55
	0.73	3.5	96	2.42	11.30	51
	0.82	3.5	66	2.68	15.76	40
B1	2.31		152	1.25	5.40	55
			128	1.50	7.25	49
			104	1.94	9.69	46
			88	2.16	12.11	42
			78	2.23	12.64	42
B2	4.60		154	1.44	6.71	51
			132	1.60	6.68	57
			100	2.02	10.46	46
			76	2.24	14.31	37
B3	9.14		250	0.72	2.85	60
			116	2.00	8.34	57

6. Annotated Bibliography for the Automatic Hydraulic Ram

6.1   "The Automatic Hydraulic Ram" - J. Krol, PROC.I.MECH.E., 1951
           vol.164, pp.103

           This paper gives a thorough analysis of the theoretical cycle
of operation of the hydraulic ram in terms of the physical dimensions
and properties of the materials making up the ram. It emphasises the
importance of the correct tuning of the impulse valve. Performance
curves for the experimental ram are given, with H_s = 13 ft (4 metres),
D = 2 inches (5 cms), hd varied, and impulse valve characteristics
varied. This is a most useful technical paper.

           "After describing the operation of a typical hydraulic ram
installation, the paper reviews the fundamentals of the water hammer as
a prerequisite to the proper understanding of the limitations of this
hydraulic machine. The historical development is discussed in some detail with the object of ascertaining what research work remained to be
done. The author presents his own theory based on the application of
general laws of mechanics to the study of a specially designed experimental hydraulic ram. By means of a theory developed, which agrees
satisfactorily with experiment, it is possible to forecast the behaviour
of any automatic hydraulic ram, provided that the following four properties at a given installation have been determined separately by experiment:
(a) loss of head in the drive pipe; (b) loss of head due to the impulse
valve; (c) drag coefficient of the impulse vales; and (d) head lost during
the period of retardation."

6.2   "The Hydraulic Ram" - N.G. Calvert, THE ENGINEER, April 19th,1957.

           Extensive experiments were carried out to understand the
performance characteristics of the hydraulic ram. This is perhaps the most def-
initive technical paper available:-

           "The possible independent variables of a hydraulic ram installation
are considered and by certain assumptions their number is reduced to eight.
Hence five dimensionless parameters are needed to describe the dependent
variables. These are the Reynolds number, the Froude number, the Mach number,
the head ratio and the coefficient of fluid friction. Each parameter is investigated in turn and it is found that the Reynolds number is ineffective
in machines of practical size and that a range exists over which the Mach
number has little influence. The Froude number is the criterion defining the
possibility of operation and (subject to a satisfactory value for the Froude

number), output and efficiency are defined by the head ratio. The optimum external conditions of operation are investigated and the conditions governing model tests are laid down".

6.3.   "Drive Pipe of Hydraulic Ram" - N.G.Calvert, THE ENGINEER
          December 26th, 1958

This paper is a continuation of the work described in reference 6.2. and gives limits to the dimensions of the drive pipe.

"In an earlier article (Calvert, 1957) the author applied the methods of dimensional analysis to a hydraulic ram installation. The relevant parameters were shown to be the head ratio, the friction coefficient, and the dimensionless numbers corresponding to those of Froude, Reynolds and Mach. Of these the first three were shown to be the most significant. In the present investigation the ram itself (as distinct from the whole installation) has been considered as an entity. The length of drive pipe then becomes an extra variable and the relevant dimensionless ratio is taken as the L/D value for the pipe. As with all the other factors connected with the hydraulic ram, knowledge of the best length of drive pipe is purely empirical. Records of systematic experiments on this variable are rare; the author does not know of any since those of Eytelwein (1803). Krol (1951) developed analytic expressions for ram performance in terms of drive pipe length and hence predicted a set of characteristic curves, but produced no experimental work in support of them."

6.4   "Hydraulic Ram as a Suction Pump" - N.G.Calvert, THE ENGINEER,
          Vol. 209, April 18th, 1960, pp.608

An adaption of the hydraulic ram to act as a suction pump is described; possible applications might include the drainage of low lying land, the emptying of canal locks, pits, etc. The hydraulic ram can be adaptable to many other uses, such as a compressor, motor, etc.

6.5   "The Hydraulic Ram for Rival Water Supply" - F.Molyneux,
          FLUID HANDLING, October 1960, pp.274

A general description of the hydraulic ram, with a design problem worked out. The impulse valve is a weighted rubber ball, which would be very difficult to tune.

6.6   "A Hydraulic Ram for Village Use" - V.I.T.A., U.S.A.

Working instructions and drawings on how to construct a small, simple hydraulic ram from commercially available water pipe fittings. The ram described had a supply head H_s = 6.5 m, delivery head h_d = 14 m, supply discharge Q_s = 35 litres/min, delivery discharge q_d = 7 litres/min. It is thus only used for small water supplies.

The impulse valve is designed to act on a sprung mechanism; the delivery valve is a simple clack valve.

6.7   "How to Design and Build a Hydraulic Ram" - Technical Bulletin,
        Technical Service Publishing Co., Chicago, 1938.

A manual on the field survey, design, construction, and installation of a simple hydraulic ram, giving step by step instructions; it includes design and performance graphs of the ram. This is a useful paper; the impulse valve of the ram appears to be unduly complicated, and the ram would not appear to be an improvement on the VITA ram.

6.8   "Rife Rams - a Manual of Information" - Rife Hydraulic Engine
        Manufacturing Co., Box 367, Millburn, New Jersey, U.S.A.

This manual refers to the rams manufactured by the company, but it gives an excellent set of instructions on the field survey, design, construction, and installation of their equipment.

6.9   "Blake Hydrams" - John Blake Ltd., PO Box 43, Royal Works,
        Accrington, Lancashire, BB5 5LP, UK.

Much of the design information in this manual used information published by J. Blake Ltd.

6.10   J. Wright Clarke - "Hydraulic Rams, their principles and
        construction". 1899 80pp
        B.T. Botsford, 94 High Holborn, London.

This technical book written in 1899 describes the techniques used in that time to construct hydraulic rams. It is interesting as it describes the effect of the ram on the inferior materials of the time and the adoptions needed to cope with the lower strengths. However, there is little information given in this publication that is not given in the other references cited.

6.11   "An Innovation in Water Ram Pumps for Domestic and
        Irrigation Use" - P.D. Stevens-Guille, APPROPRIATE
        TECHNOLOGY May 1978 vol. 5 no. 1

This article describes a hydraulic ram which incorporates two commercially available valves.

7. Appendix

<u>SOME SUGGESTIONS FOR IMPROVEMENTS TO THE RAM.</u>

1.   Received from N. Martin & R.Burton, Department of Mechanical
Engineering, Papua New Guinea University of Technology, P.O.Box 793,
Lae, Papua New Guinea.

        Experimental work on simple ram pumps is being carried out at
this University. A suggestion has been made that the elbow bend on
the ram body should be replaced by a plugged 'T' junction. The plug
can then be removed to flush out the ram without taking the impulse
valve to pieces.

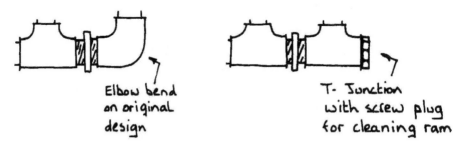

Elbow bend
on original
design

T- Junction
with screw plug
for cleaning ram

        These writers also suggest that the rubber washer for the
impulse valve is not needed, as the cushion of water flowing out of the
valve prevents hammering and wear.

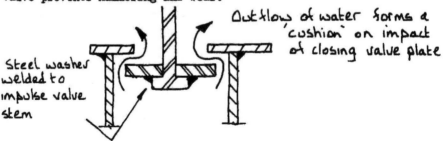

Outflow of water forms a
'cushion' on impact
of closing valve plate

Steel washer
welded to
impulse valve
stem

2.   Several different designs for the automatic hydraulic ram
pump have been tested at Eindhoven Technical College, Holland.
Details can be obtained from:-

                Appropriate Technology Unit,
                Technische Hogeschool Eindhoven,
                EINDHOVEN, HOLLAND.

3. Alternative impulse valve design sent by UNICEF Village Technology Unit, Karen Centre, near Nairobi, Kenya:

The impulse valve in the drawing overleaf has been constructed as an alternative to the leaf spring operated valve given in this manual - it has proved a reliable and satisfactory alternative.

The valve yoke is welded on to a reducing bush (3" x 2" in our case). The spring was wound from 14 gauge (2mm) fencing wire and its tension adjusted and 'locked' by the two lower nuts. The valve travel is controlled and fixed by the two upper nuts.

A further improvement would be to include a bushing (½" non-galvanised water pipe) in which the long bolt could slide, although there appears to be little friction between this bolt and the yoke. We (*UNICEF*) have therefore not incorporated it.

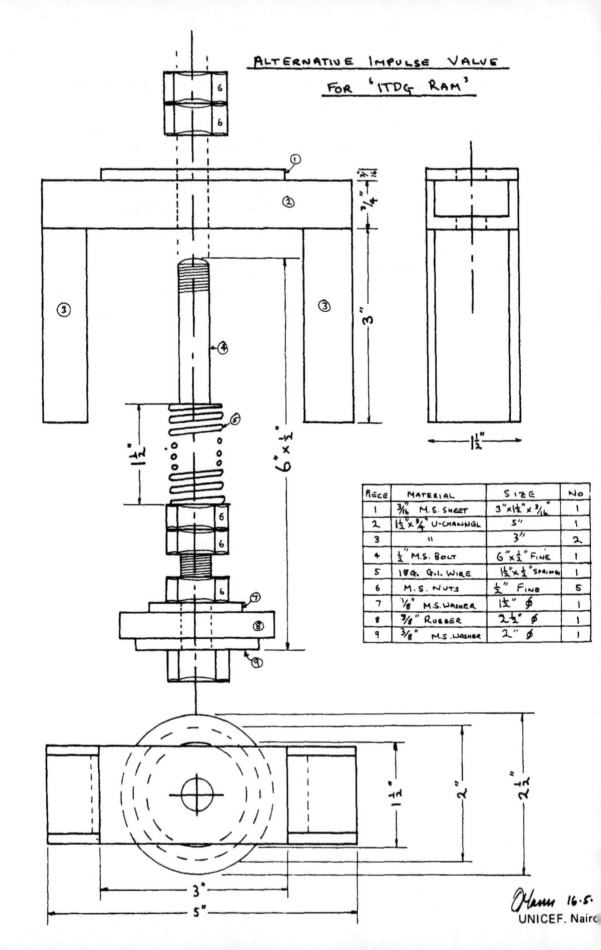

ALTERNATIVE IMPULSE VALVE
FOR 'ITDG RAM'

Piece	Material	Size	No
1	3/16" M.S. Sheet	3"×1½"×3/16"	1
2	1½"×¾" U-Channel	5"	1
3	"	3"	2
4	½" M.S. Bolt	6"×½" Fine	1
5	18g. G.I. Wire	1½"×½" Spring	1
6	M.S. Nuts	½" Fine	5
7	⅛" M.S. Washer	1½" ∅	1
8	3/8" Rubber	2½" ∅	1
9	3/8" M.S. Washer	2" ∅	1

O'Hann 16.5.
UNICEF. Nairo

www.ingramcontent.com/pod-product-compliance
Lightning Source LLC
Jackson TN
JSHW052009131224
75386JS00036B/1243

* 9 7 8 0 9 0 3 0 3 1 1 5 8 *